6th International Conference on Upstream Engineering and Flow Assurance 2018

Topical Conference at the 2018 AIChE Spring Meeting and
14th Global Congress on Process Safety

Orlando, Florida, USA
22 – 25 April 2018

ISBN: 978-1-5108-6429-0

Printed from e-media with permission by:

Curran Associates, Inc.
57 Morehouse Lane
Red Hook, NY 12571

Printed by Curran Associates, Inc. (2018)

For permission requests, please contact AIChE
at the address below.

AIChE
120 Wall Street, FL 23
New York, NY 10005-4020

Phone: (800) 242-4363
Fax: (203) 775-5177

www.aiche.org

Additional copies of this publication are available from:

Curran Associates, Inc.
57 Morehouse Lane
Red Hook, NY 12571 USA
Phone: 845-758-0400
Fax: 845-758-2633
Email: curran@proceedings.com
Web: www.proceedings.com

TABLE OF CONTENTS

Some abstracts in this book refer to images that were not made available by the publisher

515353 Effect of Electrostatic Field on Asphaltene Aggregation and Deposition Under Dynamic Conditions

Abstract:

Effect of Electrostatic Field on Asphaltene Aggregation and Deposition Under Dynamic Conditions

Vahid Azari, Erfan Abolghasemi, Shahab Ayatollahi and Vahid Taghikhani

Department of Chemical and Petroleum Engineering, Sharif University of Technology, Tehran, Iran.

Abstract

Asphaltene deposition during oil production is a leading cause in decreasing the production rate. Once occurred in an oil well, it could drastically increase the cost of remediation, workover and production. Therefore, preventative actions to control the asphaltene deposition and remediation of such heavy deposits which leads inevitably to reduction or even cessation in oil production is a main challenge. It has been shown that depending upon the nature of asphaltene molecules and the surrounding solvents, the asphaltene molecules can significantly react against electrostatic field. In other words, by applying electrostatic field precipitation, aggregation and deposition of asphaltene molecules can be controlled. To continue the previous research, in this work, the effect of electrostatic field under dynamic conditions was investigated on different crude oil samples. To do so, a homogenized mixture of crude oil and solvent was pumped into an open channel at various crude oil/asphaltene slurry flow rates using a peristaltic pump. Simultaneously, two parallel side walls of the open channel were connected to the positive and negative terminals of a DC power supply to turn them into the anode and the cathode electrodes in the electric circuit. The results showed that the asphaltene molecules, studied in this work, had tendency to be deposited on the negative electrode. Therefore, the residual surface electric charge on the colloidal molecules of asphaltene in crude oil was confirmed to be positive. Also, the effects of electric voltage, solvent/crude oil ratio and the flow rate of asphaltene/crude oil slurry in the open channel were studied on the amount of asphaltene deposition. It was shown that the electrostatic field can highly affect the aggregation rate and the deposition of asphaltene particles under flow conditions. Increasing the amount electric voltage can increase the aggregation rate of asphaltene molecules as well as and the average size of aggregates. This, in turn, can affect the rate of deposition of asphaltene molecules on the negative electrode. Also, it was concluded that an optimal crude oil flow rate was needed to be used in order to control and minimize the amount of asphaltene deposition. The results showed that the technique, proposed in this study, can be used to manage asphaltene aggregation and deposition as a preventative method. Lastly, the potential of practical application of this technique on a field scale was also discussed.

511587 Investigation of Bentonite Swelling Kinetics Using Particle Size Analysis

Abstract:

Structural fluids contain more than one phase within its solution, a liquid continuous phase with solid dispersed phase. Such fluids are widely used in most of manufacturing processes: food, pharmaceuticals, paints, plastic and solvents. In drilling of oil and gas wells, bentonite is one of the main component in formulating the drilling fluids. Bentonite dispersion in water forms a viscous structural fluid. Dispersion depends on variable factors such as particles size, shape, concentration, ion type and attraction force. Viscosity measurement for these structural fluids is critical and time-consuming process. In this paper, a low swelling Ca-bentonite is activated using soda ash combined with thermal treatment. The bentonite was successfully activated to increase and Na/Ca and hence improve the swelling capacity. Swelling of bentonite was investigated using particles size analysis. The particles size distribution distortion with time and temperature was analyzed using simple kinetic chemical reactions models. The results of particles size distribution were in agreement with viscosity measurement. The swelling kinetics of the treated bentonite fits a second â€“order model. The activation energy and the arte constants were obtained.

Acknowledgement

The authors would like to acknowledge the support of Qatar University and KFUPM for this research.

512852 Improved Oil Recovery with Viscoelastic Surfactant: Screening and Effectiveness of Formulations

Abstract:

Middle East carbonate reservoirs are usually characterized as high temperature and high saline (HTHS) reservoirs. In chemical enhanced oil recovery (cEOR), interfacial tension (IFT) and capillary number are the key parameters that alter the reservoir environment. Viscoelastic surfactant (VES) is introduced due to its dual capability of lowering the IFT and improving the sweep efficiency of the reservoir. In this work, the main objective is to screen out a novel VES for cEOR applications in HTHS carbonate reservoirs.

Novel VES solutions of 0.1 to 1 wt% concentrations were prepared with synthetic sea water and stirred until they attain equilibrium. Densities and IFTâ€™s of VES solutions and oil at different temperatures (30°C, 50°C and 80°C) were determined using density meter and spinning drop tensiometer (Kruss). Equilibrated solutions were used with heavy crude oil of 17°API to determine the IFT. Both short term and long term thermal stability of VES was evaluated using an approach based on FTIR, NMR and TGA analyses. Aging of VES was performed for 30 days at two high temperatures of 90°C and 120°C.

IFT results showed considerable reduction in IFT with novel VES and heavy oil upto 10^{-1} mN/m range. IFT decreased initially and then increased with increase in VES concentrations. Non-linear trend of IFT was obtained. IFT reduction usually depends on the concentration of solutions used. Results also indicate that ultralow IFT was obtained at 50°C. Critical micelle measurements were performed and 3wt% was found as CMC. VES was evaluated for short term thermal stability using TGA analysis and found to be stable upto 250°C. FTIR, ^{1}H NMR and ^{13}C NMR before and after aging the VES samples were compared and structural changes were observed in the aliphatic region. The peaks were shifted away due to the presence of C-O bond in the sample. IFT for aged samples was determined and almost two folds of increase in IFT was observed. Depending upon the results of this investigation at high temperatures from FTIR, ^{1}H NMR and ^{13}C NMR and IFT, this novel VES exhibit great potential to be used in cEOR applications when we have HTHS reservoirs.

513864 Protecting Environment and Assuring Smooth Energy Transfer Using Ionic Liquids

Abstract:

Gas hydrates are ice-liked crystalline compounds that are formed when small gas molecules get trapped within the water molecules under high pressure and low temperature conditions in oil and gas transmission lines. The formation of these hydrates is a major threat to oil and gas industry as they have tendency to agglomerate and completely block the oil and gas transmission lines, which may lead to an explosion or cause unwanted operations shut down. Therefore, annually industry spends around 1 billion US dollars on hydrate prevention procedures which includes extensive use of chemical inhibitors.

These chemical inhibitors are generally classified as thermodynamic hydrate inhibitors (THI) and kinetic hydrate inhibitors (KHI). The thermodynamic hydrate inhibitors function by shifting hydrate dissociation temperature to lower values and kinetic hydrate inhibitors function by delaying the hydrate formation time. The commercial THI like Methanol and Mono-ethylene glycol (MEG) perform well, but these inhibitors are required in large quantities (> 30 wt%) and cannot be easily disposed of into the environment. Therefore, there is a strong industrial need to design inhibitors that are environmentally friendly and are required in low dosage.

Ionic liquids (ILs) well known as ionic fluids are type of an organic salts that have low melting points and tendency to stay in a liquid form at low or ambient temperature. Ionic liquids are extensively being used in different chemical processes due to their negligible vapour pressure and low viscosity. Recently, ionic liquid have been recognized as the dual functional inhibitors as they have tendency to perform as kinetic hydrate inhibitor and thermodynamic hydrate inhibitor simultaneously.

In this experimental based work the thermodynamic inhibition (TI) and kinetic inhibition (KI) effect of ionic liquids (ILs) 1-Methyl-1-Propyl-pyrrolidinium Chloride [PM-Py][Cl] and 1-Methyl-1-Propyl-pyrrolidinium Triflate [PM-Py][Triflate] have been investigated on a methane rich gas mixture at different concentrations and pressure ranges (40-120 bars). The effect of addition of synergents with ionic liquids has been also studied and the experimental results have been compared with the commercial thermodynamic inhibitor methanol and literature data. All the experimental work has been conducted using PSL systemtecknik rocking cell assembly.

Acknowledgement

This work was made possible by NPRP grant # 6-330-2-140 and GSRA # 2-1-0603-14012 from the Qatar National Research Fund (a member of Qatar Foundation). The statements made herein are solely the responsibility of the authors.

512893 Flow Analysis of an Atypical Cylindrical Cyclone Separator with a Large Ratio of Length to Diameter

Abstract:

This study presents a flow analysis on an atypical cylindrical cyclone separator with a large ratio of length to diameter. Due to the distribution of the flow field determines the merits of separation performance, the aim is to optimize the flow. The Phase Doppler Particle Analyzer (PDPA) is used to measure the gas flow, and the numerical simulation is carried out by using the Reynolds Average Navier-Stokes (RANS) equation, the Reynolds stress model (RSM). Since the model is validated by good agreement between the numerical results and experimental data, and then the tangential velocity is analyzed from the transient flow field and dynamic performance.

The results showed that the tangential velocity of the transient flow field existed a non-axisymmetric phenomenon in the cross section, which was mainly presented as asymmetric distribution of the contours. The zero location of tangential velocity value was not coincide with the geometric center, where the tangential velocity was larger near the zero location. In addition, the dynamic character of tangential velocity was high-speed pulsating, with quasi-periodic characteristic. The structure or operation optimization could change the pulsation frequency and reduce the industrial vibration, which is highly significant.

Meanwhile, a parameter of RMS was used to identify the degree of turbulence velocity deviating from mean velocity. The RMS data showed that the flow instability gradually increased with the fluid flowing into the cyclone separator from the inlet section, until reached a peak value at the position about 2.0 times the length to diameter. And then the flow instability gradually decreased, with the fluid energy dissipated.

512855 Laboratory Investigation of Viscoelastic Surfactant for Chemical EOR in High Salinity and High Temperature Carbonate Reservoirs

Abstract:

Viscoelastic surfactants are gaining much attention in the recent years due to their structure. Acquiring dual capabilities of a polymer as well as a surfactant in one composition provides considerable value to chemical EOR. Middle East reservoirs are characterized as high salinity and high temperature that make carbonate reservoirs in this region even more challenging. In this work, a systematic experimental study is carried out to test the applicability of novel viscoelastic surfactant (VES-AGA 97) in enhancing the oil recovery.

Rigorous investigation of VES was performed to determine its potential in improving oil recovery. Screening of VES was carried out using five major evaluation techniques. Thermal stability of VES was determined using TGA and FTIR methods. Interfacial tension (IFT) measurements were recorded using spinning drop tensiometer. ^{1}H NMR and ^{13}C NMR were measured with long-term thermal stability tests performed for 30 days in anaerobic conditions at two different high temperatures. Static adsorption was measured by shaking the crushed powder samples with VES for 24 hours and measuring the changes in concentration of VES. Dynamic adsorption is estimated by performing core flooding experiments at 90°C.

Experimental results with TGA demonstrated that VES was thermally stable and showed high resistance to temperatures up to 250°C. FTIR results indicated good thermal stability with strong bonding between C-H and C=O. Considerable reduction in IFT with VES up to 10^{-1} mN/m range was observed. Effect of concentration, time and temperature on IFT was evaluated. It was also noted that ultralow IFT was obtained at 50°C and 80°C. ^{1}H NMR showed stability in structure while ^{13}C NMR results showed minor structural changes in the aliphatic region. Static adsorption results gave less adsorption when we had high temperatures compared to low temperatures. Langmuir and BET adsorption isotherm models were utilized to determine maximum adsorption rate. Comparison between both models and experimental results exhibited perfect match. Dynamic adsorption was evaluated using concentration analysis of effluents collected at the downstream from core flooding. Difference between the concentrations of VES at inlet and outlet was measured using total organic carbon analysis. Additional oil recovery of 8% was determined when recovery factor is plotted against pore volumes.

512730 Investigation on the Nature of Asphaltene Flocs

Abstract:

Asphaltene flocculation is an important step in some settling processes involving solvent diluted bitumen and is a contributing step to asphaltene deposition in pipelines and oil processing facilities. Asphaltene flocculation occurs when asphaltenes precipitate from an oil upon depressurization or the addition of a poor solvent. Asphaltene precipitation is considered by some to be a colloidal flocculation process (Eyssautier *et al.*, 2012) and by others to be a phase transition (Johnston *et al.*, 2017; Dini *et al.*, 2016). Once an oil is destabilized or undergoes a phase transition, asphaltenes grow from nano-aggregates of a few nanometers in diameter to particles of approximately 1 micrometer in diameter which in turn form flocs that are hundreds of micrometers in diameter (Angle *et al.*, 2006). These flocs are assumed to be held together by electrostatic and van der Waals forces and therefore their size distribution is expected to be shear sensitive. Population balance models have been used to model the kinetics of asphaltene flocculation based on the assumption that there is a dynamic equilibrium between flocculation and shattering as found in other chemical systems (Maqbool *et al.*, 2011). The objectives of this study are to determine if the underlying assumptions of the proposed flocculation process are in fact correct and to propose an appropriate model for the observed behavior.

Asphaltene flocs were examined from two Western Canadian bitumen samples diluted with *n*-pentane or *n*-heptane at concentrations above the onset of precipitation. The asphaltene floc size distribution was measured over time in a series of batch experiments at different diluent contents, shear rates, and contact times. All experiments were carried out at room temperature and atmospheric pressure. The floc size distributions were measured while mixing using a focused beam reflectance method apparatus. These measurements were validated with micrographic images of samples collected *in situ*. The fractal dimension of the flocs was determined from the volume of the settled flocs.

In all cases, the asphaltenes formed particle (or floc) size distributions with number average diameters of tens to hundreds of micrometers. The distributions were established in less time than the first measurement could be obtained (about 30 seconds) and only changed slowly, if at all, afterwards. The initial floc size distribution was shear dependent. The average floc size increased with diluent content but reached a plateau value at 75 to 85 wt% diluent.

Micrographic images and fractal dimensions indicated that, near the onset of precipitation, compact linear and planar floc structures dominated. At higher diluent contents up to approximately 85 wt% diluent the flocs remained compact but became more three-dimensional. Above 85 wt% diluent, the flocs became more porous and branched. The fractal dimensions increased from approximately 2.3 near the onset to approximately 2.8 as the diluent content increased to 80 wt% and then decreased to 2.2 at higher diluent contents. Higher fractal dimensions indicate more spherical and/or more compact flocs. Near the onset of precipitation, no significant change in floc size was observed over time (up to 5 hours). At dilutions above 80 wt% solvent, the average floc size decreased with shear while the number of flocs increased, consistent with shattering. No re-flocculation was observed at any conditions even when the shear rate was reduced.

We hypothesize that the flocs first form via a rapid nucleation and growth process that can be modeled as flocculation. The asphaltenes are sticky in this phase and the flocs are fused structures. We further hypothesize that the asphaltenes rapidly lose their stickiness such that no further flocculation occurs. After the first few seconds, only shattering can occur. The compact structures formed near the onset are difficult to shatter and therefore shear insensitive while the looser structures formed at high dilution can be partially shattered under the shear conditions of these experiments. Further tests to validate these hypotheses will be discussed. A shear-dependent Smoluchowski aggregation model is proposed for the steep growth of the flocs in the first seconds. A second shattering-only model is proposed to account for the effect of shear over time.

Keywords: asphaltene, flocculation, shattering, bitumen, fractal dimension, population balance.

References

Angle, C.W., Long, Y., Hamza, H., Lue, L. Precipitation of asphaltenes from solvent-diluted heavy oil and thermodynamic properties of solvent-diluted heavy oil solutions. *Fuel*, 85, 2006, 492-506.

Dini, Y., Becerra, M., Shaw, J.M. Phase behavior and thermophysical properties of Peace River bitumen + propane mixtures from 303 K to 393 K. *Journal of Chemical Engineering Data*, 61, 2016, 2659-2668.

Eyssautier, J., Frot, D., Barré, L. Structure and Dynamics of Colloidal Asphaltene Aggregates. *Langmuir*, 28, 2012, 11997-12004.

Johnston, K., Schoeggl, F.F., Satyro, M.A., Taylor, S.D., Yarranton, H.W. Phase Behavior of Bitumen and *n*-Pentane, *Fluid Phase Equilibria*, 442, 2017, 1-19.

Maqbool, T., Raha, S., Hoepfner, M.P., Fogler, H.S. Modeling the aggregation of asphaltene nanoaggregates in crude oil-precipitant systems. *Energy & Fuels*, 25, 2011, 1585-1596.

512884 Viscosity Modeling of Visbroken Heavy Oils

Abstract:

Heavy oil viscosity exceeds 1000 mPa.s and bitumen viscosity can be as high as 1 million mPa.s at atmospheric conditions. The viscosity of these fluids must be reduced prior to transport and processing and one option to do so is visbreaking. Visbreaking is a mild liquid-phase thermal cracking process where large hydrocarbon molecules react in order to produce shorter chain molecules [1]. Visbreaking is optimized by modifying the process severity (residence time and temperature) to maximize the production of distillates and viscosity reduction in the residue, while avoiding coke formation [1]. Hence, to design and optimize a visbreaking process, it is necessary to predict the composition and viscosity of the reacted fluid. The overall objective of this study is to develop a method to predict the viscosity of the visbroken crude oil from the feed crude oil properties and reaction conditions. The method is to be structured for implementation in process simulators.

It has been shown that the viscosity of visbroken products cannot be predicted from the properties of the whole product oil [2] but there is potential to predict the viscosity from the changes in the distribution and properties of the oilâ€™s pseudo-components. The proposed methodology is to characterize the oil into a set of pseudo-components that represent the property distributions within the fluid. Each pseudo-component is assigned the set of properties necessary to model the reactions and calculate viscosity. These properties, including molecular weight, density, and boiling point, are determined from a distillation assay, a SARA (saturate, aromatic, resins, asphaltene) assay, and established correlations. The viscosity is to be determined with the Expanded Fluid (EF) viscosity model [3]. The model inputs for the feed can be determined from the molecular weight, specific gravity, and boiling point. To complete the model, a method is required to predict the mass fraction and model parameters of the distillable and SARA pseudo-components of the visbroken products. The specific objective of this contribution is to develop correlations for the model parameters of the visbroken product.

A Western Canadian bitumen sample was visbroken at five different residence times and temperatures in an in-house laboratory pilot plant. In addition, a feedstock and two thermally cracked samples were provided by a Chinese industrial source. The pilot plant is a continuous flow coil visbreaking unit. The feed oil is pumped through a pre-heater at a temperature of 250°C and then to a reactor where it reaches the desired visbreaking temperature. The reactor is held at sufficient pressure to maintain a single liquid phase in the reaction zone. The product from the reactor enters an atmospheric flash separator. The separated gas is vented while the liquid product is collected for further analysis.

The feed and visbroken product were fractionated using distillation and SARA assays. The conversion was determined based on the change in the mass fraction of the 524+°C cut in the product versus the feed. The viscosity, density, and molecular weight of these fractions were measured. The dataset from the Western Canadian bitumen and its visbroken products was used to develop correlations for the Expanded Fluid parameters for the distillables and each SARA fraction based on reaction conditions. The viscosity of the visbroken products was then predicted from the product oil characterization and the correlated parameters. The methodology was then used to predict the viscosity of the visbroken Chinese oil samples.

References:

[1] Joshi, J; Pandit, A; Kataria, K; Kulkarni, R; Sawarkar, A; Tandon, D; Ram, Y; Kumar, M. Petroleum residue upgradation via visbreaking: A review. *Ind. Eng. Chem.* 2008, **47**, 8960-8988.

[2] Rueda-Velazquez, R; Gray, M. A viscosity-conversion model for thermal cracking of heavy oils. *Fuel*. 2017, **197**, 82-90.

[3] Ramos-Pallares, F; Taylor, S; Satyro, M; Marriot, R; Yarranton, H. Prediction of viscosity for characterized oils and their fractions using the expanded fluid model. *Energy. Fuels*. 2016, **30**, 7134-7157.

513163 Application of Aggregation Thermodynamics and Molecular Characterization to Study Asphaltene Precipitation of Bituminous Crude Oils

Abstract:

Adding paraffinic feedstock to bituminous crude oil significantly increases the risk of asphaltene precipitation. Both the onset point and yield of asphaltene precipitation are determined by the content and composition of the crude mixture. Comprehensive thermodynamic modeling explains and enables the prediction of such precipitation behavior. Recently, an aggregation thermodynamic framework [1] developed from Yen-Mullins hierarchy model for asphaltene [2] has been shown to successfully predict asphaltene precipitation from a wide range of solvent mixtures, which is applied in this study. Additionally, a molecule-based characterization technique [3] is employed to identify the molecular compositions of the petroleum fluids. Finally, the solvent power of the crude oil and its blends is calculated using a refined NRTL-SAC activity coefficient model. To demonstrate this methodology, we predict the asphaltene precipitation behavior of three heavy oils: Cold Lake, Athabasca, and Peace River.

Reference

[1] M. Wang, Y. Hao, M. R. Islam, and C.-C. Chen, "Aggregation thermodynamics for asphaltene precipitation," AIChE Journal, vol. 62, pp. 1254-1264, 2016.

[2] O. C. Mullins, "The Modified Yen Modelâ€ ," Energy & Fuels, vol. 24, pp. 2179-2207, 2010.

[3] C.-C. Chen and H. Que, "Method of Characterizing Chemical Composition of Crude Oil For Petroleum Processing," US Patent Application Number: 2013/0185044, July 18, 2013.

513567 Ultra Small-Angle X-Ray Scattering (USAXS) Assessment of Asphaltene Precipitation and Inhibitor Effectiveness

Abstract:

The mechanisms of asphaltene precipitation and inhibition remain unsolved despite nearly a century of focused investigation. Previous characterization approaches either study the flocculation and dynamics of asphaltenes >500 nm in size with optical microscopy or on length scales <50 nm using X-ray and neutron scattering approaches. Ultra-Small-Angle X-ray Scattering (USAXS) is a technique that has not been previously applied to precipitating asphaltene systems and provides valuable insight into the structure of soft systems on length scales from 1 nm to approx. 5 microns. The experimental results to be presented demonstrate the unique capabilities of USAXS to measure time-resolved values of size, molecular weight, and internal structure (i.e., fractal dimension) of precipitating asphaltenes. The results demonstrate that USAXS is a valuable new tool to assess asphaltene phase behavior. The experimental results have numerous applications in aggregation model validation, fundamental mechanistic insight, and inhibitor screening. Asphaltene precipitation with and without chemical inhibitors (nonylphenol and nonylphenol formaldehyde resin) were monitored by adding n-heptane to a toluene-diluted vacuum residue sample. Both of the inhibitors screened demonstrated an ability to slow the precipitation growth rate of the asphaltene clusters. The results demonstrate that one of the inhibitors (nonylphenol) more effectively reduces the rate of asphaltene precipitation. The experimental results provided by USAXS go beyond assessing the rate of growth and can also provide valuable insight into the mechanisms of inhibition. The internal structure of precipitating asphaltene clusters or aggregates can also be assessed with USAXS to monitor morphological changes as a result of solvent or inhibitor addition, providing valuable design and mechanistic insight.

512339 Using Temperature Measurements from Production Logging/Downhole Sensors to Diagnose Multistage Fractured Well Flow Profile

Abstract:

The temperature log as a part of a production logging package has been used in the industry for many years. Recent developments in downhole fiber optic sensing technology introduced a more continuous measurement of temperature both temporally and spatially for flow diagnosis. One of the powerful applications of temperature measurements is to diagnose multistage fracture treatments. Because the fiber optic temperature measurements can be available during fracturing, during shut-in, and during production, the integrated interpretation provides information of fracture/flow distribution. This information helps to understand what happened during the fracture treatment stages, to identify problems in treatment design and execution, and to improve the efficiency of multistage fracture stimulation.

The key component of fracturing diagnosis by temperature measurements is the interpretation models. In this paper, we will introduce the models developed for the purpose of fracture diagnosis. The mathematical models are built on mass, momentum and energy balances of each component (reservoir, fracture, well completion and wellbore) in the system, and the components are linked through the boundary conditions. All models can be solved numerically, but for computational efficiency, analytical/semi-analytical solutions are preferred when available. To correctly simulate heat transfer during fracture propagation, a fracture geometry model with appropriate leakoff description is integrated in the model buildup. For flow problem in a fractured well, analytical model, streamline approach and reservoir simulation can all be the solution methods. The paper compares the advantages of each approach.

Field cases from the Eagle Ford, the Marcellus, and the Vaca Muerta shale formations will be presented in the paper to illustrate how the models can be used to generate the fracture/flow distribution. The results show that temperature measurement is a comprehensive tool for fracture diagnosis.

516267 A New Two Part Friction Reducer System Enables Lower FR Concentration and Allows up to 100% Reuse of Produced Water â€“ Mechanisms and Case Study from the Marcellus Shale

Abstract:

Friction reducers (FRs) are an important component of slickwater hydraulic fracturing applications. Using only a single FR system throughout the entire treatment is highly desired in water sources of salinity varying up to 300,000 ppm. This paper discusses the working mechanisms and field trials of a new salt-tolerant FR system in the Marcellus shale.

A three-well trial program was initiated in the Marcellus. Multiple water sources with varying salinities were used with up to 100% reuse of produced water. The new FR system enabled the pumping pressure to be reduced below 8,500 psi, and the pumping rates were increased and maintained at approximately 100 bbl/min.

This new salt-tolerant FR system consists of a water-in-oil cationic polymer and a surfactant inverter. Unlike other FRs, the distinctive advantage of this FR is that the ratio between polymer and inverter can be readily adjusted on-the-fly to achieve maximum friction reduction with reduced friction reducer concentration.

513338 Mobility Improvement Using Water Soluble Polymers: Role of Chemical Structure and Anionicity

Abstract:

Acrylamide-based polymers such as partially hydrolyzed polyacrylamide (HPAM) and copolymer of acrylamide and 2-acrylamido-2-methylpropane sulfonic acid (AMPS-AM) are being used a mobility control in chemical EOR. The rheology of their solutions has been studied extensively in several conditions such as increasing molecular weight, salinity, and temperature. Most of the previous study mainly deals with the evaluation of polymers at reservoir conditions without having fundamental insight. In this work, we investigated how the rheology of a polymer changes by changing the polymer structure, anionicity, and degree of hydrolysis.

Steady and dynamic shear rheology of eight polymers from HPAM and AM-AMPS group were investigated at 25°C and 80°C using high-pressure high-temperature Discovery hybrid rheometer. These polymers contain different molecular weight, different functional group, and varying degree of hydrolysis. In steady shear rheology, the viscosity was measured at a shear rate ranging from 0.1/s to 1000/s. The viscoelastic properties such as elastic storage and loss modulus were determined by dynamic shear rheology by varying frequency from 0.1 rad/s to 100 rad/s.

The anionicity and degree of hydrolysis found to have a major impact on the rheological properties of the polymers. This result was consistent for both HPAM and AM-AMPS copolymers both at high and low temperature. The viscosity increased by increasing the anionicity. The elastic storage modulus also increases by increasing the molecular weight and anionicity. Molecular weight and anionicity also influence the steepness of viscosity reduction with temperature and the change in energy dissipation capacity with time at high temperatures. The possible influences of charge distribution and the nature of the anions present in the polymer need to be considered when considering the type of polymer (HPAM of AM-AMPS copolymer) for a specific EOR application.

This study brings the novel insight of polymer structure-property relationship that can help in quickly screening the polymers for EOR application in carbonate reservoirs.

513684 Effect of Crude Oil on the Reaction Kinetics of Diesel Emulsified Acid with Calcite at High-Pressure and High-Temperature HPHT

Abstract:

Emulsified acids have been used as an effective well stimulation fluid since 1933. It is primarily an emulsion solution consisting of two immiscible liquids, one of them is dispersed (inner or dispersed phase) in the other liquid (continuous or outer phase) in the form of droplets. It used in oil fields because it has several advantages. Firstly, its used to minimize the danger of corrosion of the tubular system including a casing, tubing, etc. Secondly, it has good reaction retardation properties that lead to deep penetration into the formation and finally the high viscosity of the emulsified acid allows it to sweep the targeted region and distribute uniformly. In order to model the carbonate acidizing treatment, there is an important experimental parameter that should be estimated which is the diffusion coefficient of H^+ (De) to the surface of the reservoir rock. The diffusion coefficient is calculated by measuring the reaction rate at different RPM using rotating disk apparatus RDA. Almost, all the diffusion coefficients values calculated from RDA experiments for calcite with diesel emulsified was done through water-saturated disk at lower pressure that is not enough to keep CO2 in solution. However, this work, study the effect of crude oil saturation as well as HPHT condition on the diffusivity of H^+ to the surface of the limestone, which mimics the real reservoir condition where emulsified acids are used.

In this work, two sets of reaction kinetic experiment were conducted using rotating disk apparatus RDA. The first set of oil-saturated disks, and the second for dry calcite ones. Emulsified acid was prepared using acid to diesel ratio of 70:30, emulsifier concentration and corrosion inhibitor were used in 1 vol.% and 0.3 vol. % respectively. Compositional analysis for diesel was done. Limestone core samples from local outcrop were used after checking its purity using XRD and XRF analysis. A small disk of 1.5 in. the diameter and 0.75 in. the length was prepared. The saturation by crude oil was made by soaking the disks in crude oil for 48 hours at 65â,˚C. All the rotating desk experiments were conducted at 3000psi and 135â,˚C. The disk rotational speed varied between (200-1500 rpm). Effluent samples of 5 ml were taken every two minutes and then analyzed through Inductively coupled plasma mass spectrometry (ICP-MS). The dissolution rate was measured at each RPM an

From the lab results a couple of two conclusions were drawn, firstly the Diffusion coefficient of H^+(De) value at HPHT is much lower than those obtained at low-pressure low- temperature LPLT and the extrapolation of these De values at lower downhole conditions for deep reservoirs could lead to severe acidizing and hydraulic fracturing design inaccuracy. Secondly, two De values were observed for oil saturated limestone reaction kinetics, one for lower RPM where crude oil saturation effect is highly impacting the diffusion and leading to lower reaction, whereas at higher ones the (De) value is almost close to the (De) value for dry calcite rock.

To sum up, from lab results and for better acidizing modeling Itâ€™s recommended to use diffusion coefficient (De) calculated using disks saturated with reservoir fluids at downhole conditions.

A Critical Overview of Cause and Detection of Pipeline Blockages

Lu Yang, Huirong Liang, Alexey Kovalev, Sai Wang, Kegang Ling,
University of North Dakota, Grand Forks, ND

Abstract

Due to dramatic energy renaissance in US since 2010, more pipelines carrying crude oil and petroleum products are constructed to meet the gathering and transportation requirements. Blockage, however, can occur anywhere because of the depositions of wax, hydrate, paraffins and sand in pipes, and seriously affects the safety and reliability of transportation process. Therefore, it is imperative to detect the location and size of blockage in pipelines more accurately and efficiently to reduce the number of pipeline accidents.

Before any physical detections is taken, it is important to make an appropriate assessment of the nature and severity of blockages, which can be mitigated or removed to avoid possible pipeline incident. Then selected inspection methods can provide accurate location and size of blockages. It is known that physical methods such as acoustic reflectometry, gamma-ray transmission scanning, radioisotope tracer injection, tomography measurement, radiographic detection, and pipeline diameter measurement can detect blockage accurately but with high cost/time consuming and necessary equipment or materials, while mathematical models used to detect blockage are at a relatively low cost and can monitor the pipeline continuously without interrupting pipeline operations but with less accuracy. Therefore, understanding of cause of blockage and the advantage/disadvantage of each detection method are crucial to selecting an appropriate method to detect blockage in a specific pipeline with a relatively low cost while could still meet the flow assurance requirement.

In this study, blockage nature is assessed in several aspects including fluid characteristic, blockage varieties and formation mechanism. Upon obtaining initial assessment of the blockage, the suitability of blockage detection method is studied for different fluids in a pipe of various layouts and operating conditions. By comparing the advantages and disadvantages of difference detection methods, a systematic instruction is proposed to choose the right method for differentiating single or multiple blockages in the pipeline and then estimating the location and size of blockage in different scenarios. Finally, it is found that mathematical method has the privilege to narrow down the possible blockage interval especially for long-distance pipeline, then physical method is used to locate blockage accurately and evaluate severity confidently.

Early information of pipeline blockage can improve the effectiveness and reliability of pipelines operating. This paper highlights the importance of assessing the nature of blockages correctly and focuses on developing best practice to detect the pipeline blockages in different scenarios. Furthermore, for the comparison of numerous detection methods, this study addresses a practical and systematic guideline that instructs how to select efficient and appropriate detection methods, which also allow the operator to assess the risks and schedule remediation procedures efficiently and economically.

Keywords: pipeline blockage, physical inspection, mathematical model

Introduction

Pipeline blockage, which is resulted from deposition caused by changes in pressure, temperature, or composition, is a critical issue in oil & gas production and transportation systems and is affecting oil companies and customers worldwide. It is therefore no surprise that blockage has attracted higher attention by researchers over the past years. Especially as the boom of the shale reservoirs production, more new pipelines are constructed to meet the transportation requirement of ever-increasing crude oil and natural gas, refined products, carbon dioxide, and production water. At the same time, the original pipe networks are aging. New pipelines as well as the original pipe networks will inevitably experience more blockage problems during operation, which also increases the need for more intelligent blockage detection methods to better evaluate and locate blockage occurs.

Some detection methods have been proposed to characterize pipeline blockages. Acoustic signals (Wu, 1994) measured from the pipeline was proposed to detect blockage by analyzing the signal frequency shifts. Since the quick decay of the acoustic signals, the pipeline interval can be detected is limited to hundred meters. Detection method was carried out by using ROV (remotely operated vehicle) (Rogers, 1995) to measure the pipeline blockage, while the response from the blockage is slow. Scott and Yi (1999) proposed that the transients flow were affected by the blockage location but failed to localize the position. Another method is the pressure transient (Adewumi, et al., 2000; Ahmed, 1996) carried out by initiating the pressure wave at the inlet, which can be used to detect the single blockage in the pipeline. They also proposed a mathematical model that propagating the pressure pulse through a gas pipeline with multiple blockages and demonstrated the possibility to locate multiple blockages by analyzing the pressure waves at the pipeline inlet. Chen et al. (2007) applied the theory by Adewumi et al. (2003) to a laboratory experimental setup, they found the blockage location and length were estimated accurately except for the blockage severity. Improvements in computer memory and speed also encourage researchers concentrate on developing more accurate mathematical models to characterize the pipeline blockages.

Several papers have reviewed pipeline blockage modeling and management. A review and classification of single and multiple blockages detection methods are summarized by Wang (2004) and Adewumi (2003). Another review of simple numerical modeling is given by Adeleke (2012). A summary report summarizing specific detection procedures about blockage can be found in Sampath K. (2016). Unlike the existing reviews mentioned above, which are focusing on a particular blockage aspect (usually blockage detection), this paper looks wider by considering the overall blockage management process. The objective of this paper is to review the detection methods and models proposed in the past in different phases, from analyzing the blockage characterization to estimating the location and severity of the blockage. It is expected that in this way the new systemic guideline will be proposed for different scenarios. The main objective is to identify the advantage and disadvantage of all detection methods and then use the guideline to decide the appropriate approach to detect the pipeline blockage.

This paper is laid as follows. After this introduction, the relevant background information about blockage is introduced in section 2. This is followed by the review of blockage detection methods in section 3, and systemic guideline for choosing the appropriate detection method in section 4. Finally, the main conclusion is drawn in section 6 of this paper.

Blockage characterization

Pipeline blockage can be caused by a number of different factors. Some examples include the formation such as asphaltene, wax, or natural gas hydrate because of critical operating conditions like high pressure or low temperature. Blockage also can form by sand deposition or scaling. Due to the incompatibility of chemical or well streams. Other common factors that generate blockage are mechanical causes like stalled pigs, collapsed pipe or failed valves. Due to its variety formation conditions, pipeline blockage can and does occur anywhere from the sand surface to the export pipe networks (Bukkaraju, 2016). From the field operation and flow assurance engineering experience, some of the following events suggest the partial pipeline blockage:

- Increased pressure drop along the flow path from the inlet to the outlet.
- Reduced flow rate suddenly or complete loss of the fluid.
- Increased difference in temperature along the pipeline from original value because of the cooling or heating effect.
- Presence of small hydrate, wax presence, or sand in the routine pigging returns.

Existence of a pipeline blockage poses serious risks. Pipeline blockage can not only reduce the operation efficiency of pipe network, but also cause a safety problem if the blockage has not been detected accurately and removed quickly. The accident can reduce the profit of project or delay the project significantly (Rui et al., 2017a, 2017b, and 2017c). It is believed that accurate location of pipeline blockage can reduce the risks by appropriate detection methods. While before any detection method is taken, it is necessary to make appropriate assessments of the nature and severity of the pipeline blockage. Blockage characterization is assessed in several aspects including fluid characterization, operating condition and pipe system assessment.

(1) Fluid characterization

Understanding the produced fluid characteristics and the propensity to fluid mechanisms, are of critical importance to estimate the potential risk of generating blockage in pipeline system. Pipeline blockage varies due to the difference between the fluids being transported. The major threats for flow assurance are proposed by Sampath K. (2016) that include hydrate, wax and asphaltenes. Hydrate blockage, always found in gas pipelines, has not only high dependency on low temperature or high pressure, but also on water presence. Wax blockage in crude oil pipeline is formed when the fluid flow cools down, especially wax crystals are usually transported and deposited on the pipe wall during winter. Asphaltenes presence in reservoir fluid are usually dependent on system pressure changes, with the risk of plugging the reservoir perforations. Fluid compositional analysis (e.g. water-cut, gravity, GOR), fluid properties analysis (e.g. viscosity, dew point), hydrate/ wax/ asphaltene characterization, fluid assurance state analysis (e.g. steady state or transient state), and fluid mechanics can be of benefit for blockage characterization.

(2) Operating condition

Tracking the pipeline system operating parameters helps greatly in assessing the nature of the blockage. Simulation and fluid modeling also help estimate the fluid characteristics. Operating pressure, temperature and fluid conditions along the pipeline system need to be known in order to effectively estimate with higher accuracy the nature of the blockage. High pressure and low temperature driven the pipeline blockages such as hydrate/ wax/ asphaltens may occur simultaneously at some certain conditions, but can require different methods to detect and estimate. Besides, under some operating conditions not only the partial blockage is present, but some multiphase blockages or multiple blockages will be present especially for the pipeline over long distance. Flow assurance and operation experience have shown that in many field cases once the operating conditions are misunderstood, wrong methods for blockage detection and remediation are undertaken subsequently. This may be further increasing the risk of accidents and cost because of exceeding operating pressure.

(3) Pipe system assessment

Fluid phase change is not the only cause of blockage formation, the mechanical failures of various components of pipe systems can also lead to pipeline blockage. Understanding the pipe system is critical important to detect the potential blockage locations. Estimating the possibility of blockage formation includes several factors like pipeline layout and all fitting distributions. Flow back points, stalled pigs, defective valves, umbilical lead connections, or ruptures in the internal pipeline system are the typical locations that the blockages can form. On the other hand, most of the blockages can form again although they have been mechanically removed, it is of great importance to study historical operating data such as daily production rates, operating conditions, and prior blockage intervals which are variables at the time of blockage formation. Hydrates in gas pipeline can form and deposit in hours, for instance, while the wax or asphaltenes blockage would take several weeks to form under the corresponding operating conditions.

Pipeline blockage can be partial or full based on its cause and the nature of the blockage. The appropriate assessment of the nature of pipeline blockage is ascertained through fluid characterization, which can help select the proper detection method to estimate the location and severity of the pipeline blockage.

Blockage detection methods

An accurate estimation including the location, size and severity of pipeline blockage would lead to savings in cost and time for the operator. Although some blockage detection methods have been around for years, with continuing development, they are becoming high tech and sophisticated. Based on the detection equipment needed or not, the detection methods developed so far can be broadly classified into the following two main groups: (a) physical inspection methods and (b) mathematical models.

3.1 Physical inspection methods

The objective of physical inspection methods is to estimate and locate the blockage by corresponding detection equipment, primarily including isotope tracking inspection, densitometry measurement, and acoustic reflectometry.

(1) Isotope tracking inspection

According to the fluid continuity equation, the area with higher velocity indicates restriction. This technology involves the injection of an isotope into the pipeline and subsequently their movements are tracked using the detecting device. The operating data recorded by the tracking device actually indicate the velocity of the isotope along with the fluid which in return indicates the possible location and diameter of the blockage. This isotope method was firstly used by MacIntyre (1959), Wagner et al. (1961) to visualize the cardiac blood pool. After decade's development, the isotope inspection method has become popular tools in industry. Charlton et al. (1981) helped find the outstanding isotope that much easier to be tracked for the detecting device. Kasban (2010) also applied the isotope method for detecting blockage or leakage along the pipeline.

(2) Densitometry method

Base on the assessment of the blockage characterization, the densities of blockage formed by hydrate, wax, asphaltene, or scales are of great difference between each other. Densitometry method detects the density of the fluid at some pipeline intervals or cross sections by the specific scanning device. The differences in the densities along the flow path indicates the possible location of the blockage. While the differences in densities between the blockage and the transported fluid also affect the accuracy of this detection method. For example, the density of hydrate blockage is close to the density of water which may be difficult to differentiate these two kinds of blockages (Detta et al., 2016). However, engineering field experience coupled with the information about the fluid property and operating parameters would help in estimating the existence and location of the blockage.

(3) Acoustic reflectometry

This technology involves the introduction of a pulse of sound in to the pipeline by the acoustic pulse generator. When the acoustic pulse encounters the blockage, reflection is produced and passes along the pipe. The pulse signal is measured by acoustic sensors that are installed outside of the pipe, and then the information can be used to plot a noise profile along the whole pipe. Deviations from the baseline profile indicate the location of pipeline blockage. Theoretical and experimental work about acoustic reflectometry was first proposed by Parker (1981) who studied the correlation between acoustic pulse and background noise, hence developed approach that very small acoustic signals to noise ratio could be detected even for long-distance pipeline. Acoustic techniques were originally used to leakage detection in pipeline (Watanabe et. al. 1986, 1987a, 1987b, 1990), Koyama et al. (1993) applied acoustic technique for the location and severity of blockage in the pipeline. Wang et al. (2009) performed an experiment in 16m long PVC pipeline by acoustic methods to detect blockage. The reflected acoustic signals were recorded using matched filters to overcome the influence of background noise.

Apart from these detection techniques, other physical inspection methods such as diameter measurement, radar and sonar technology, thermography method and gamma ray scanning are also used in the detection of the blockage. Physical inspection method can estimate the existence and locate the position of blockage directly and accurately by detecting devices while in return it increases the operating cost significantly at the expense of shutting down and cannot monitor the pipeline network operating continuously.

3.2 Mathematical models

Mathematical model utilizes the mass conservation, momentum conservation, energy balance equation coupled with the operating parameters such as temperature, pressure, flow rate to estimate the existence and location of blockage. Compared with physical inspection methods, mathematical models have the advantage of quick evaluation at lower cost and can monitor the pipeline continuously without interrupting pipeline operations especially in some harsh environments like deep water, polar area or areas with higher difficulty in accessing. Mathematical models primarily include backpressure technology, pressure transient pulse technology, frequency response method and dimensionless method.

(1) Backpressure technology

This method involves a multi-flowrate experiment undertaken which establishes a baseline profile of pressure drop versus flow rate. Deviations from this profile indicates the possible existence and location of blockage. This technology has long been recognized as an efficient method for detecting blockages in pipeline. Scott and Satterwhite (1998) considered the application of backpressure method as a method monitoring the growth of blockage in gas pipeline and first proposed the blockage factor to estimate the location of blockage. Scott and Yi (1999) also applied this method in liquid flow lines while this method only helps in rough estimation because the effects of length and size of the blockage are coupled together. Liu and Scott (2000,

2001) proposed the average pressure method on the basis of backpressure technology to locate the partial blockage in the pipeline. This new method requires three different tests: steady-state back pressure test that is identical to backpressure method taken to determine the blockage factor; simultaneous shut-in test to obtain the average pressure; bleed-off test to determine the volume factor that indicates the location of pipeline blockage.

(2) Pressure transient technology

This technology involves the generation and transmission of pressure pulse through the liquid medium along the pipeline. The time delay between the pressure pulse and the echoes received by the topside receiver is used to locate the position of blockage with reasonable precision. Vitkovsky (2003) sent an impulse generated by the movement of the valve down the pipeline to detect the location of blockage which also can be used for leakage detection. Adewumi et al. (2003) proposed a one-dimensional model based on the pressure transient method to describe the propagation of a pressure pulse through a pipe under different scenarios like no blockage, single blockage and multiple blockages cases. Chen et al. (2007) proposed a 1D representation model to investigate the characteristics of the pressure wave propagation process through a pipe with blockage, besides, this model can be applied to estimate the blockage location, length and severity in single or multiphase pipeline. Adeleke et al. (2012) proposed a single phase, isothermal gas model taking viscous effects into account when detecting blockage. The viscous losses have no effect on blockage length and location prediction accuracy but has significant impact on the accuracy of blockage severity predictions.

(3) Frequency response method

This method involves the introduction of pressure pulse generated by a valve opened and closed periodically. The procedure is repeated for a range of frequency and the amplitude of the pressure fluctuations at the location of the oscillating valve is analyzed by using the transfer matrix method. The comparison between peak pressure frequency responses indicates the location and size of partial blockage. A frictionless model based on frequency response was proposed by Mohapatra (2006) to characterize the location and the size of blockage. Especially when the pre- and post-frequency development of a pipeline is known, this model can be used to detect the location and the size of additional partial blockage. Lee et al. (2008) found that discrete blockages were influenced by the frequency peaks on an oscillatory pattern by analyzing the behavior of the pipeline in the form of a frequency response diagram. He also pointed out this method can be extended to situations with unknown operating parameters. Duan et al. (2011) proposed a detection model of blockages in water pipeline by analyzing the occurrence of the resonant peaks in the frequency axis.

The merits and demerits of these above mentioned methods are also discussed in Table 1. The comparison of applicability of different methods is given in Table 2.

Table 1. Merits and demerits of different detection methods

Detection methods		Merits	Demerits
Physical Inspection methods	Isotope tracking inspection	• Be effective in detecting the locations and sensitivity of blockages in partially blocked pipelines.	• Be more challenging based on the pipeline design, i.e. pipe in pipe or for buried pipelines are similar to those noted for the densitometry technique.
	Densitometry method	• This method can be repeated multiple times to predict the blockage location with an increased level of confidence.	• Be more challenging based on the pipeline design, i.e. pipe in pipe, buried pipes. • The density of some of the typical pipeline contents are very close.
	Acoustic reflectometry	• Methodology is economical, proficient to identify blockages and holes in pipe as small as 1% of its diameter	• To monitor longer pipelines, a large number of acoustic sensors are needed.

			• Small blockages whose acoustic signal is small and only differ slightly from the background noise cannot be detected.
Mathematical models	Backpressure technology	• This method can be used to provide important information for monitoring chemical inhibition programs, scheduling pigging, assessing risk for pigging and planning other intervention procedures.	• The higher pressure drop is more preferable to a lower overall pressure drop for locating the blockage because with limitation of pressure gauge the location result may be questionable. • The subtle difficulty in using this method lies in the establishment of the baseline, which requires a time-consuming multi-rate test.
	Pressure transient technology	• This method is remote, non-intrusive and efficient. • No additional instrument other than a dynamic pressure gauge is needed.	• In the case of hydrate blockages, the addition of liquid head and pressure pulsing may aggravate the blockage situation. • The blockage severity is usually underestimated.
	Frequency response method	• The flow variables at only one location are sufficient to predict the blockage. • The methodology is simple and is economical. • The PPFR can be used to estimate the location and the size of the blockage.	• Decision on the range of frequencies should take into account the safety of the system and constraints on valve operation. • Valve operation continuous opening and closing may require specialized instrumentation. • The fluctuation of valve opening should be kept low, say about 0.1 so that the linearity assumption is not violated.

Table 2. Applicability of different methods

Detection methods	Type of fluid			Nature of blockage	Blockage Diagnosis		
	High density(e.g. Water)	Medium density(e.g. Oil)	Low density(e.g. Gas)		Location	length	diameter
Isotope tracking inspection	√	√	√	All	√	√	
Densitometry method	√	√	√	All	√	√	√
Acoustic reflectometry	√	√	√	All	√	√	
Backpressure technology	√	√	√	All	√	√	
Pressure transient technology	√	√	√	Non-hydrate blockages	√	√	√
Frequency response method	√	√	√	All including Single and two blockages	√	√	

Guideline

Early assessment of pipeline blockage can improve the effectives and the reliability of pipeline operation. Based on the comparison of different detection methods, a systematic and practical guideline for selecting the appropriate and efficient detection method is proposed as follow:

(1) Pipeline blockage is characterized in several aspects including fluid properties, operating conditions and pipe system assessment, which is of importance to make an appropriate assessment of the nature and severity of the blockage.

(2) According to pipeline operating parameters like flowrate, temperature, pressure, and pipe length, several mathematical models are applied to evaluate the possible location and length of pipeline blockages. Especially the number estimation about blockages along the pipeline is of great importance for blockage detection accuracy.

(3) To compare the results of mathematical models, several dimensionless variables (Ling et al., 2016) are introduced. Dimensionless length: the pipe length is turned into dimensionless variable whose domain is from 0 to 1 (measured from the inlet to the outlet of the pipeline). Dimensionless blockage location: the ratio of possible blockage location (measured from the inlet of the pipeline to the blockage location) to the total length of the pipe. Dimensionless blockage length: the ratio of possible blockage length to the total length of the pipe.

(4) Plot dimensionless location curves based on data obtained in step 1 and 2 as shown in Figure 1. In this plot, x-axis is dimensionless blockage location and length, y-axis is different mathematical models, and z-axis is dimensionless blockage diameter.

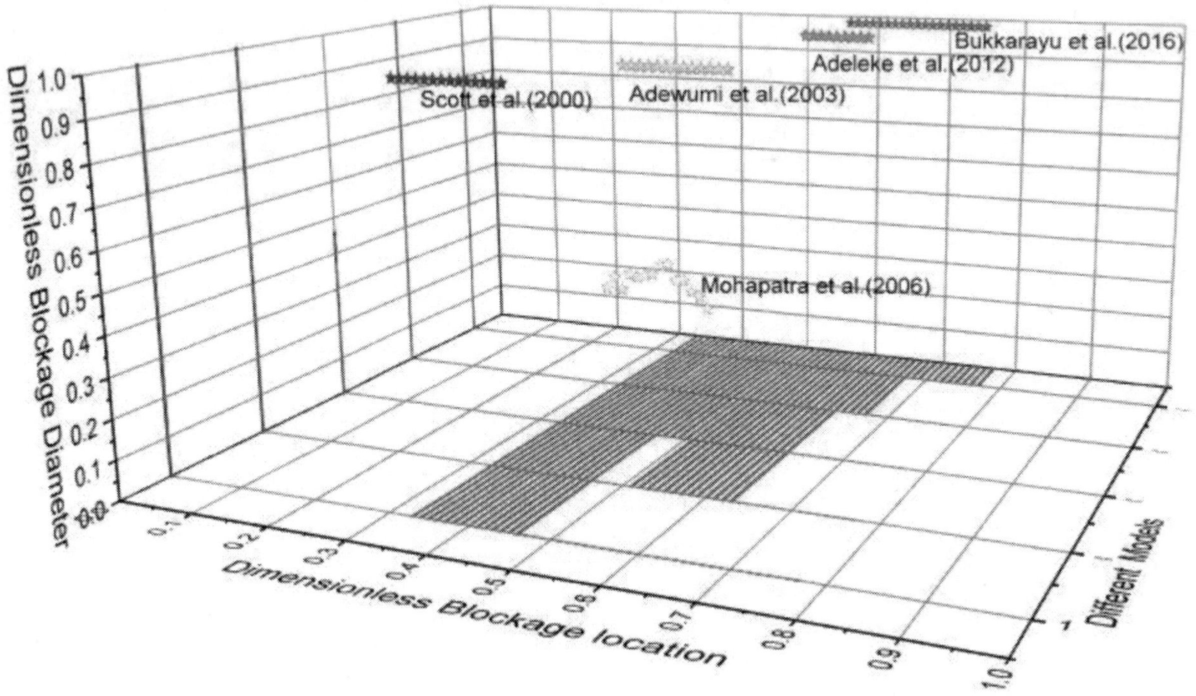

Note: Assuming the dimensionless blockage diameter of 1 (full blockage) when plotting the curves because some of mathematical models are unable to evaluate the diameter of the blockage.

Fig.1- Plot the dimensionless curves for different mathematical models

(5) Find the overlap part of dimensionless blockage location which is the most possible blockage location, calculate the possible blockage location. For example, the overlap part of dimensionless pipe length in Fig. 1 is from 0.32 to 0.75.

(6) According the dimensionless blockage length obtained from the over part dimensionless blockage location, calculate the possible blockage length.

(7) Once obtained the possible location and length of the pipe, then the physical inspection method chosen based on the assessment of blockage is applied to evaluate accurately the diameter of the blockage.

Pipeline blockages and leak can occur anywhere in the pipeline and is of higher difficulty to be detected accurately in the pipe over long distance because one or more blockages and leaks can occur (Rui et al., 2017). Under the direction of this systemic detection guideline, the mathematical models are implemented to get the first evaluation of pipeline blockage like location and length with much lower cost compared with physical inspection method. Within all those possible locations of blockage, then the physical inspection methods are applied to detect the location, length and diameter of the blockage with higher accuracy. By following this guideline, the inspection length could be narrowed down largely with mathematical models that based on the operating parameters. The guideline also enable the operator to assess the risks and schedule remediation procedures efficiently and economically.

Conclusions and future work recommendations

Blockage characterization, detection methods have come a long way. Based on the review completed and presented here, following main conclusions and future work recommendation are made.

Before any detection methods are implemented, it is important to assess what is contributing to the blockage formation. Blockage nature is assessed in several aspects including fluid characteristic, blockage varieties and pipe system.

Upon obtaining the initial assessment of pipeline blockage, the merits and applicability of blockage detection method are studied for different fluids in various operating conditions. By comparing the difference of detection methods, a systematic instruction is proposed to choose the right method for estimating the location and size of blockage in different scenarios.

It is found that mathematical method has the privilege to narrow down the possible blockage interval especially for long-distance pipeline, then physical method is applied to locate blockage accurately and evaluate severity confidently. Attention needs to be taken at this stage to make sure that the detection method itself would not deteriorate the blockage or make it harder to remediate in future work.

With more reliable, timely and cost-saving blockage detection methods, more accidents causing the damage to the facility, possible loss of life, and environmental pollution can be avoided. Many mathematical models are available to detect pipeline blockages while their applicability is limited to single or two blockages in a pipeline, which just covers a small part of pipeline operation circumstances. The future work of detection methods should focus on how to detect the multiple or multiphase blockages in more complicate pipelines like parallel/looped pipelines.

Acknowledgment

The authors are grateful to The Petroleum Engineering Department at the University of North Dakota. This research is supported in part by the Research North Dakota Venture Grant Phase I under award number 17-05-J1-169 as well the Intercollegiate Academics Funding from the office, Vice President for Research & Economic Development.

Reference

1. Adeleke, Najeem, Mku Thaddeus Ityokumbul, and Michael Adewumi. "Blockage detection and characterization in natural gas pipelines by transient pressure-wave reflection analysis." SPE Journal 18.02 (2012): 355-365.

2. Adewumi, Michael A., E. S. Eltohami, and A. Solaja. "Possible detection of multiple blockages using transients." Journal of energy resources technology 125.2 (2003): 154-159.
3. Ahmed, W. H., 1996, ''Location of Partial Blockage in Natural Gas Pipelines Using Transient Modeling,'' M.S. Thesis, The Pennsylvania State University, University Park, PA.
4. Bukkaraju, Sampath K., et al. "Practical Guidelines for the Diagnosis and Remediation of Pipeline Blockages." Offshore Technology Conference. Offshore Technology Conference, 2016.
5. Charlton, J. S., J. A. Heslop, and P. Johnson. "Radioisotope techniques for the investigation of process problems in the chemical industry." Industrial application of radioisotopes and radiation technology: proceedings of an International Conference on Industrial Application of Radioisotopes and Radiation Technology. Vol. 598. Intl Atomic Energy Agency, 1982.
6. Chen, Xianghui, et al. "Pressure-wave propagation technique for blockage detection in subsea flowlines." SPE Annual Technical Conference and Exhibition. Society of Petroleum Engineers, 2007.
7. Datta, Shantanu, and Shibayan Sarkar. "A review on different pipeline fault detection methods." Journal of Loss Prevention in the Process Industries 41 (2016): 97-106.
8. Duan, Huan-Feng, et al. "Extended blockage detection in pipelines by using the system frequency response analysis." Journal of Water Resources Planning and Management 138.1 (2011): 55-62.
9. Kasban, H., O. Zahran, and FE Abd El-Samie. "New trends for on-line troubleshooting in industrial problems using radioisotopes." Journal of Electronics and Electrical Engineering 2.3 (2010): 284-292.
10. Koyama, H.,Watanabe, K. and Himmelblau, D. M., Identi_cation of Partial Plugging in Gas- Transport Pipeline by ab Acoustic Method, Applied Acoustics, Elservier Science Publishers Ltd, England, Vol. 40, 1-19, 1993.
11. Ling, Kegang, Xingru Wu, and Zheng Shen. "A New Method to Detect Partial Blockage in Gas Pipelines." Oil and Gas Facilities 5.05 (2016).
12. Liu, Liang J., and Stuart L. Scott. "A new method to locate partial blockages in subsea flowlines." SPE Annual Technical Conference and Exhibition. Society of Petroleum Engineers, 2001.
13. Liu, Liangjian, and Stuart L. Scott. "Development of a type curve to locate partial blockages in gas flowlines." SPE Annual Technical Conference and Exhibition. Society of Petroleum Engineers, 2000.
14. MacIntyre, William J., et al. "The visualization of internal organs by accentuation scintillation scanning technics." Radiology 73.3 (1959): 329-340.
15. Mohapatra, P. K., et al. "Detection of partial blockage in single pipelines." Journal of Hydraulic Engineering 132.2 (2006): 200-206.
16. Parker, John, Acoustic Detection and Location of Leaks in Underground Natural Gas Distribution Lines", John Hopkins APL Technical Digest, Vol. 2, pp. 90-101, 1981.
17. Qunli, Wu. "Reconstruction of blockage in a duct from single spectrum." Applied Acoustics 41.3 (1994): 229-236.
18. Rogers, L. M. "Pipeline blockage location by strain measurement using an ROV." Offshore Technology Conference. Offshore Technology Conference, 1995.

19. Rui, Z., Han, G., Zhang, H.,Wang S., Pu, H., Ling, K., 2017. A new model to evaluate two leak points in a gas pipeline, Journal of Natural Gas Science and Engineering, 46, 491-497.
20. Rui, Z., Lu, J., Zhang, Z.,Guo, R., Ling, K., Zhang, R., Patil, S., 2017. A quantitative oil and gas reservoir evaluation system for development, Journal of Natural Gas Science and Engineering, 42, 31-39.
21. Rui, Z. Li, C., Peng, P., Ling, K., Chen, G., Zhou, X., and Chang, H., 2017. Development of Industry performance metrics for offshore oil and gas project, Journal of Natural Gas Science and Engineering, 39, 44-53.
22. Rui, Z., Peng, F., Chang, H., Ling, K., Chen, G., and Zhou, X., 2017, Investigation into the performance of oil and gas projects, Journal of Natural Gas Science and Engineering, 38, 12-20.
23. Scott, S. L., and L. A. Satterwhite. "Evaluation of the backpressure technique for blockage detection in gas flowlines." Journal of energy resources technology 120.1 (1998): 27-31.
24. Scott, S. L., and Yi, J. (1999). "Flow testing methods to detect and characterize partial blockages in looped subsea flowlines." Journal of Energy Resources Technology, ASME, 121, 154-160.
25. Vitkovsky, J. P., et al. "Leak and blockage detection in pipelines via an impulse response method." Pumps, electromechanical devices and systems applied to urban water management 1 (2003): 423-430.
26. Wang, Xiao-Jian, Martin F. Lambert, and Angus R. Simpson. "Detection and location of a partial blockage in pipeline systems using damping of fluid transients." Critical Transitions in Water and Environmental Resources Management. 2004. 1-10.
27. Wang, X., Lennox, B., Turner, J., Lewis, K., Ding, Z., 2009. Blockage detection in long lengths of pipeline using a new acoustic method. In: 16th International Congress on Sound and Vibration, Krakow, Poland, pp. 1e8.
28. Wanger, Henry N., John G. Mcafee, and James M. Mozley. "Diagnosis of pericardial effusion by radioisotope scanning." Archives of internal medicine 108.5 (1961): 679-684.
29. Watanabe, K. and Himmelblau, D. M., Detection and Location of a Leak in a Gas- transport pipeline by a new acoustic method," AIChE J., Vol 32, No. 10, pp.1690-1701, 1986.
30. Watanabe, K. Koyama, H. and Ohno, H.,Location of a leaks in gas - transport pipeline by the acoustic method," Proc. Of International Soc. Of America (ISA)/87. Int. Conf. and Exch., ISA, pp. 619-26, 1987.

513573 Kinetic Precipitation of Asphaltenes with the Presence of Inorganic Solids

Abstract:

This work investigated the influence of inorganic solids on the rate of kinetic precipitation of asphaltenes from diluted bitumen. At a fixed amount of precipitated asphaltenes, higher solid contents significantly enhanced the rate of asphaltenes precipitation, resulting in no significant observable kinetics under certain conditions. The results suggested a heterogeneous precipitation mechanism for asphaltenes exists when inorganic solids are present in bitumen or petroleum samples. A combined homogeneous aggregation and heterogeneous nucleation model was developed to analyze the results. The rate of heterogeneous nucleation was observed to be upwards of 90% faster than the rate of homogeneous aggregation, but as the quantity of precipitated asphaltenes increased, the rate of homogeneous aggregation could exceed that of heterogeneous nucleation. Scanning Electron Microscope with Energy-Dispersive X-ray Spectroscopy was used to investigate the inorganic solid properties. The results suggest that kaolinite is likely the mineral present in the studied samples that exhibit the heterogeneous precipitation mechanism; however, other materials may also behave similarly. This investigation leads to a better understanding of asphaltene aggregation in systems with inorganics solids and provides a simple model that can estimate the rate of asphaltene precipitation under different conditions. Moreover, this work provides a potential new tool to increase the rate of the often-slow kinetic precipitation of asphaltenes using inorganic solids as nucleation sites to reduce experimental time.

513865 Probing the Effect of Stirring on Natural Gas Hydrate Formation-a Future Energy Fuel

Abstract:

The transport of natural gas through liquefaction is one of the major modes of natural gas export across different regions in the world. The process of liquefaction of natural gas is costly and consume large sum of energy as it requires cooling of up to -161°C. Normally, the liquefaction cost vary from 3 $ to 10 $ per MMBtu (millions of British thermal units). So if the market prices of crude oil are down, the liquefaction cost could negatively affect LNG business and make liquefaction projects unprofitable.

In order to avoid liquefaction cost, the scientists are trying to study gas hydrates as potential mode of natural gas transportation and storage in future. Gas hydrates are crystalline compounds, formed of water and gas molecules. One volume of hydrate can store up to 180 volumes of natural gas at STP (standard temperature and pressure). Therefore, the development of hydrate production plants holds particular importance in future.

In this experimental based work, the formation rate of gas hydrates in a high pressure cell has been studied at different stirring rates using a magnetic stirrer. In the first stage, the system was first cooled from 20 °C to 2 °C at constant pressure for 18 hours. Then in second stage, the system was left stable at 2 °C for another 18 hours at constant stirring rate. The stirring rates were varied between 100-1400 Rpm (Rotations per minute) for each experiment and the gas hydrate crystal formation was observed by calculating pressure drop across the high pressure cell during the second stage.

The experimental result shows that for a chosen high pressure cell, the stirring rate has a significant effect on the rate of gas hydrate crystal formation. The maximum gas hydrate crystal formation was observed within the stirring range of 550-750 RPM. It was found that there exists a threshold stirring limit, above or below which little or no hydrate crystal formation takes place. This fundamental study is a significant step forward and it can be useful for modeling purposes and designing of a pilot scale hydrate production plant to study the economical and safety factors associated with natural gas hydrate storage and transportation phenomena.

Acknowledgement

This work was made possible by NPRP grant # 6-330-2-140 and GSRA # 2-1-0603-14012 from the Qatar National Research Fund (a member of Qatar Foundation). The statements made herein are solely the responsibility of the authors.

513871 Inhibitory Action of Cost-Effective Ionic Liquids Methane Hydrates

Abstract:

Inhibitory Action of Cost-Effective Ionic Liquids Methane Hydrates

Tausif Altamash[a] , M. Fahed Qureshi[a,], Majeda Khraisheh[a*] and Mert Atilhan[b]

[a]Department of Chemical Engineering, Qatar University, P.O.Box 2713, Doha, Qatar.

[b] Department of Chemical Engineering , Texas A&M University, College Station, TX, USA.

The gas hydrate formation is the most common experience during transportation of natural gas from one place to another place. Actually, transportation takes place through pipeline under high pressure and low temperature and this thermodynamic condition trigger the hydrate formation. Initially formation of hydrate started from encapsulation of gas molecule by water molecules as particle which turn into large size and block the pipeline at the end. In order to avoid this shutdown process chemical inhibitors like Methanol and Methyl ethylene glycol or salts came into light, but they give unavoidable environmental and mechanical risks. Therefore, to overcome those issues ILs being used since recent years as greener low dosage inhibitor to protect environmental hazards. ILs said to be a dual nature (thermodynamic and kinetic) designer gas hydrate inhibitors, nevertheless the cost of ILs can not be ignored. In the search of cost effective ILs, protic ionic liquids (PILs) can be tested to use up to larger dosage. PILs are another class of ionic liquids (ILs) that having great opportunity to fit in multidisciplinary application, especially in electrochemistry. PILs are synthesized on proton transfer phenomena, on addition of BrÃ¸nsted acid and BrÃ¸nsted base. Chemically, the presence of acceptor-donor site provides a hydrogen bond to build it up and this easy rout of reaction cut preparation cost effectively. However, in the presence of water PILs strengthen the ionâ'ion water-PIL interaction. The affinity of PILs with water provide an advantage to use them as gas hydrate inhibitor which may disrupt the water structure and thus slow down the process of gas hydrate formation up to some extent. in this study very first ever PILs [EA][Of], [DMA][Of] and [DMEA][Of]) [TBMA][Of]) were used at 5 and 10 wt.% respectively to investigate the thermodynamic and kinetic inhibitions at different ascending pressure conditions using the rocking cell assembly (RC-5) for methane gas. Investigations reaves that HVLE curve was shifted by 2.0 ± 0.1 °C and hydrate formation was delayed by 2 hrs at high pressure conditions (~120 bars). However, kinetic inhibition also performed well and enhanced the time of gas hydrate formation. Moreover, poly ethylene oxide (PEO) were added to examine the synergistic behaviour of duo (PILs+PEO)

AcknowledgementThis work was made possible by NPRP grant for # 6-330-2-140 (TA) and GSRA # 2-1-0603-14012 (FQ) for from the Qatar National Research Fund (a member of Qatar Foundation). The statements made herein are solely the responsibility of the authors.

* To whom correspondence should be addressed. **E-mail:** m.Khraisheh@qu.edu.qa

AUTHOR INDEX